Grade 3 · Unit 3

Inspire
Science

Different Environments

Mc
Graw
Hill
Education

FRONT COVER: (t)Andrea Izzotti/Shutterstock, (b)Dr Morley Read/Shutterstock

Mheducation.com/prek-12

STEM McGraw-Hill is committed to providing
instructional materials in Science, Technology,
Engineering, and Mathematics (STEM) that give all
students a solid foundation, one that prepares them
for college and careers in the 21st century.

Send all inquiries to:
McGraw-Hill Education
8787 Orion Place
Columbus, OH 43240

ISBN: 978-0-07-699628-5
MHID: 0-07-699628-X

Printed in the United States of America.

4 5 6 7 8 9 10 11 LKV 26 25 24 23 22 21 20

Survive the Environment

Change of Environments

Survive the Environment

How do animals survive in this environment?

Hunting

⟳ GO ONLINE
Check out *Hunting* to see the phenomenon in action.

💬 Talk About It

Look at the photo and watch *Hunting*. What questions do you have about the phenomenon? Talk about your observations with a partner.

Did You Know?

Coyotes can run as fast as 40 miles per hour. They also walk on their toes to avoid being detected by predators.

Design an Animal's Adaptations

How do some organisms survive in some environments but others cannot? You are being hired as an animal curator. At the end of the module, you will develop a design for a newly discovered animal in a specific habitat. Your goal will be to design and build a model of the animal in its habitat.

What do you think you need to know before you can design a new animal's adaptations?

Animal curators manage exhibits in aquariums and zoos. They know a lot about the animals and what they need to survive in their environment.

HIRO
Ocean Engineer

STEM Module Project

Plan and Complete the Science Challenge Use what you learn about adaptations to reach your goal.

PAGE KEELEY
SCIENCE
PROBES

Will the Animals Survive?

There is a type of animal that lives in the forest. These animals eat tree nuts and seeds. They find shelter high up in the treetops. Their fur helps them blend in with the bark of the tree. What would happen to these animals if all the trees in the forest were cut down? Circle the answer that best matches your thinking.

A. *They would all die.*

B. *Some would survive if they were born with a characteristic that would help them live in the changed environment.*

C. *Some would survive by changing their food, shelter, or fur so they could live in the changed environment.*

Explain your thinking. Describe what happens to organisms when their environment changes.

_____

_____

_____

You will revisit the Page Keeley Science Probe later in the lesson.

Survival of Organisms

ENCOUNTER
THE PHENOMENON

How does the tree survive?

GO ONLINE

Check out *Bristlecone Pine Trees* to see the phenomenon in action.

Talk About It

Look at the photo of the bristlecone pine tree and explore the *Bristlecone Pine Trees* digital activity. What questions and observations do you have? Talk about them with a partner.

Did You Know?

The oldest bristlecone pine tree is more than 5,000 years old!

INQUIRY ACTIVITY

Hands On

Plant Hunt

Explore different locations of your schoolyard
and find the same plant in each of those locations.

Make a Prediction Do plants of the same type have
different characteristics in different locations?

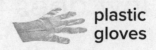
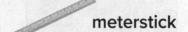
_____

_____

_____

Carry Out an Investigation

BE CAREFUL Wear gloves when handling plants.

1. Walk around your schoolyard. Find the same
   plant growing in two different locations.

2. **Record Data** Observe each plant.
   Count the number of leaves and place
   in the table.

	Location 1	Location 2
Number of leaves		
Plant height		
Width of largest leaf		

3. Measure the height of the plant and the width of the
   largest leaf. Record the measurements in the table.

Communicate Information

4. Was there a difference in the plants from each location?

_____

_____

_____

_____

_____

5. Did your results support your prediction? Explain.

_____

_____

_____

_____

_____

_____

💬 **Talk About It**

Why do you think the plants had different measurements?

Plant Needs

Like all organisms, plants have needs. Plants get everything they need from their environment. Living things that do not get what they need may die. Plants need air, water, nutrients, light, and space to live.

Plants get a gas called carbon dioxide from their environment. Plants take in the gas through their leaves. They use this gas to make food.

Like all living things, plants need water. They take in water through their roots. Water travels from the roots, up the stem, to the leaves. Plants use water for many life functions. Water helps a plant to stand up. It keeps a plant from wilting. A plant also uses water to make food.

Nutrients are substances that help living things grow and stay healthy. Nutrients are dissolved in water. Plants absorb nutrients when they take in water through their roots.

The green parts of plants collect the energy in light and use it to make their own food.

Plants need space to grow and to get water and sunlight. Different plants need different amounts of space.

Although most plants grow in soil, a plant can also grow in water without soil if it gets the nutrients it needs.

1. Look back in the text and circle five things plants need in order to live.

2. Think back to the Inquiry Activity, *Plant Hunt*. Why might the plant in one location grow taller and greener than the plant in the other location?

_____

_____

_____

Copyright © McGraw-Hill Education   Pixtal/age fotostock

INQUIRY ACTIVITY

Materials

Hands On

Needs of Plants

You observed the same plant in different environments. Investigate how plants change to meet their needs. Compare the growth of two plants.

Make a Prediction What will happen to a plant if obstacles block its direct light?

 prepared shoebox

 2 bean plants

 cup of water

 ruler

_____

_____

_____

Carry Out an Investigation

1. **Record Data** Use the ruler to measure the height and width of the plants. Record your data in the table on the next page. Include any observations you make about the plants.

2. Place one plant in the box prepared by your teacher, on the end opposite of the opening. Leave the other plant outside of the box.

3. Put the lid on the box. Turn the box so the opening is facing toward a bright light source. Place the unboxed plant next to the box.

4. Record your measurements and observations in the data table on the next page on days 5, 10, and 15. After recording your measurements on each of those days, give your plants some water and replace the lid. Return the plants to their original positions if you moved them.

INQUIRY ACTIVITY

Unboxed Plant

Day	Plant Height (cm)	Observations
0		
5		
10		
15		

Boxed Plant

Day	Plant Height (cm)	Observations
0		
5		
10		
15		

Communicate Information

5. Describe the growth of both plants over time.

_____

_____

_____

_____

6. Did your findings support your prediction? Explain.

_____

_____

_____

Animal Needs

In order to stay alive, animals need certain things. These include food, water, oxygen, space, and shelter.

Animals need food because it gives them energy to move and grow. Different animals have different physical traits to help them get food. Some meat eaters, like tigers, have sharp teeth to help them bite and tear meat. Many plant eaters, such as cows, have large, flat teeth for chewing.

Animals need water because it helps them turn food into energy and get rid of waste.

Animals need oxygen, which is a gas. Animals get oxygen by breathing. Most land animals use lungs to get oxygen from the air. Some animals that live in water get oxygen from the water by using gills.

Animals need space to move around, grow, find food, and raise their young. Different animals need different amounts of space.

Animals need shelter, or a safe place to be. No animal can be alert all the time, which means it needs somewhere safe to go. Zebras live in herds, so some zebras can keep watch while others sleep.

Elephants use their trunks to lift drinking water to their mouths.

When in danger, a turtle will hide its head, legs, and tail inside its shell.

1. Circle five things animals need to survive.

2. What would happen if a tiger did not have sharp teeth?

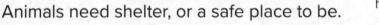

Ecosystem

Living things live in ecosystems. An **ecosystem** includes all the living and nonliving things that interact in an environment. Living things include plants and animals. Nonliving things include rocks, soil, water, and air.

A **resource** is a material that living things use to survive. Living things get resources from their ecosystems. For example, plants and animals get the air and water they need from their environments. But every ecosystem has a limited amount of resources. As a result, living things must compete for them. **Competition** is the struggle among living things for resources. When organisms cannot compete, they cannot get the resources they need. They may die or move to another ecosystem.

 Explain what could happen to the **ecosystem** below if the cattails disappeared.

A Pond Ecosystem

Crane flies eat plants and algae. They lay eggs in water.

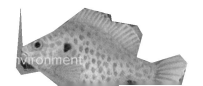

WRITING Connection Write a paragraph about an animal's habitat from the animal's point of view.

GO ONLINE Explore *Animal Fun Facts* to learn about different animals.

_____

_____

_____

_____

_____

_____

_____

REVISIT
PAGE KEELEY
SCIENCE
PROBES

Revisit the Page Keeley Science Probe on page 5.

Cattails grow well in wet soil. Some animals use them as food and shelter.

...rtles climb out of the water to warm up in the sunlight.

Pond snails slide along the bottom, looking for plants and algae to eat.

Copyright © McGraw-Hill Education

Inspect

Read the passage *Birds*. Underline text that tells what the word *drought* means.

Find Evidence

Reread How did the drought change the Pacific Flyway habitats? Highlight text evidence that supports your answer.

Notes

_____

_____

_____

_____

_____

_____

_____

Hooded orioles can be found in many neighborhoods in the western United States.

Birds

Many birds are native to North America. The hooded oriole and the mourning dove are two examples of North American birds. These birds are able to survive in their habitats because they can meet their needs. They like to live in open woods along streams. The hooded oriole especially likes palm trees.

From 2011 to 2017, there was a drought in the southwest region of the United States. There was not enough water, so many plants died. This affected all of the wildlife, because many animals are herbivores. Herbivores eat only plants. When the plants died, these animals could not find enough to eat.

Copyright © McGraw-Hill Education Fuse/Corbis/Getty Images

Mourning doves can be found from southern Canada to central Mexico.

Make Connections
💬 Talk About It
Discuss with a partner how a drought in parts of the Pacific Flyway could affect the birds in this passage.

Notes

Birds also depend on water. The drought changed much of the area known as the Pacific Flyway. Birds migrating south from Alaska depend on habitats in the southwestern United States for feeding and regaining their strength before continuing their journey south. With dried up wetlands and streams along migration routes, some birds had trouble surviving. Dried up wetlands meant there were fewer fish and insects too.

During periods of change in these habitats, birds may fly elsewhere to find what they need. This means that there are more birds in fewer places. More birds in fewer habitats could lead to fewer resources and more competition. How do you think this would affect the hooded oriole and mourning dove?

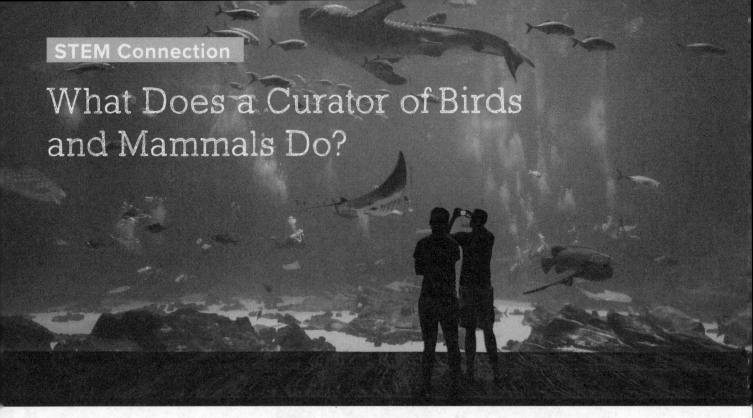

What Does a Curator of Birds and Mammals Do?

Describe your career.

I work directly with all kinds of marine mammals and birds. I help design animal habitats. I work with my staff and our veterinarian to create the best environment for each type of animal we have. I deal with the health issues of our mammals and birds on a daily basis.

How did you get interested in this career?

I was born and raised in Long Beach, California, and was always outside. I spent much of my time around water and at the beach. I was a good swimmer. I took SCUBA diving lessons at the youngest age possible, thirteen. My favorite classes in school were science and math, so I naturally wanted to take those courses in college. This led me to study zoology while at university. I use my education in my job every day.

What's an everyday problem you see and could solve?

I work to make sure the water quality of our marine exhibits is excellent every day. In the past, aquariums were found on the coasts. They could bring fresh ocean water in to the exhibits and then send it back out to the ocean.

As humans became more aware of the impact of the environment and pollution, we had to make changes. Aquariums on the coast and those inland moved to a closed water system. A closed water system means aquariums could not release waste to the ocean anymore. We had to solve the problem of how to offer the best environment for our animals. We use a lot of science and math to make sure our water is clean and healthy for the animals. The water must also be clear enough for visitors to see. This is something I deal with and solve on a daily basis.

Dudley is the Curator of Birds and Mammals at an aquarium in Long Beach, California.

Describe an accomplishment that you are proud of.

The aquarium likes to educate people about what is happening in our oceans. One way of letting people know is to exhibit animals that are strongly affected by changes in their environment. One accomplishment that I am proud of is helping with the design of our Magellanic penguin exhibit. I helped design the habitat. I also helped bring the penguins to the aquarium. Many of the penguins were stranded. The exhibit is now complete. The birds are comfortable in their safe surroundings.

What additional questions would you have for a curator of birds and mammals?

_____

_____

_____

_____

EXPLAIN
THE PHENOMENON

How does the tree survive?

Summarize It

Explain how organisms survive in their environments.

_____

_____

_____

_____

_____

_____

_____

REVISIT
PAGE KEELEY SCIENCE PROBES

Revisit the Page Keeley Science Probe on page 5. Has your thinking changed? If so, explain how it has changed.

Three-Dimensional Thinking

1. Which of the following is not a need for plants?

    A. water

    B. sunlight

    C. shelter

    D. space

2. What happens if an organism does not get the resources it needs?

3. Describe how animals experience competition in an ecosystem.

Extend It

Think about your school environment. What items are essential to your success and growth as a student? Create a brochure for a school that only has these essential items. Discuss with your classmates whether you would prefer attending a school that only has your essential items rather than your current school.

_____

_____

_____

_____

_____

_____

_____

_____

_____

_____

KEEP PLANNING
STEM Module Project
Science Challenge

Now that you have learned what organisms need to survive, go to your Module Project to explain how the information will affect how you design your model.

Adaptations

Polar bears live in the cold Arctic. They grow a coat of thick fur to stay warm.

Two friends were at a zoo in San Diego. The zoo had a polar bear exhibit. They wondered how the polar bear could live in San Diego where it is very warm. This is what they said:

Suzanne: *The polar bear will try to adapt by growing less fur.*

Milo: *The polar bear will not try to adapt by growing less fur.*

Who do you agree with the most? _____

Explain why you agree.

_____

_____

_____

You will revisit the Page Keeley Science Probe later in the lesson.

Adaptations and Variations

ENCOUNTER
THE PHENOMENON

How does the quail survive in its environment?

◐ GO ONLINE

Check out *Quail* to see the phenomenon in action.

💬 Talk About It

Look at the photo and watch the video of the quail. What questions do you have about the phenomenon? Talk about your observations with a partner. Record your questions below.

Did You Know?

During a drought, quails may get their water from insects and succulent vegetation.

INQUIRY ACTIVITY

Hands On

Bird Beak Shapes

Plants and animals have traits that help them survive in their environments. Observe how birds with different types of beaks gather food by using a model of the beaks.

Make a Prediction How does the shape of a bird's beak affect its ability to gather different types of food?

_____

_____

_____

_____

Materials

 safety goggles

 dried macaroni

 paper clips

 paper plates

 plastic cups

 plastic spoon

 rice

 toothpicks

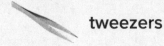 tweezers

Carry Out an Investigation

1. Collect a plastic spoon, tweezers, and a toothpick. These will represent the birds' beaks.

2. Place paper clips, rice, and macaroni each on a separate plate in front of you. These will represent bird food.

3. Place a plastic cup next to each of the plates. Choose a bird beak model.

4. Your teacher will set a timer for 20 seconds. When the timer starts, use one bird beak model to collect as much of one type of food as possible. Collect food by using the beak to move the food into the cup.

Copyright © McGraw-Hill Education   (2)olgaman/iStock/Getty Images, (3)McGraw-Hill Education, (1 4 6)Ken Cavanagh/McGraw-Hill Education, (others)Jacques Cornell/McGraw-Hill Education

5. Count the amount of food you were able to collect. Record your data in the table below.

6. Repeat steps 4 and 5 until you have collected each food type with each beak model.

Amount of Food Collected			
Model Beak	Paper Clips	Rice	Macaroni
Spoon			
Tweezers			
Toothpick			

Copyright © McGraw-Hill Education

INQUIRY ACTIVITY

Communicate Information

7. Which bird beak worked best for each type of food?
Use evidence to support your ideas.

_____

_____

_____

_____

_____

8. Compare your results with your classmate's results. Are they
the same? What do you think it might mean if the results
were different?

_____

_____

_____

_____

9. Did your results support your prediction? Explain.

_____

_____

_____

MAKE YOUR CLAIM

How do animal features help them survive?

Make your claim. Use your investigation.

CLAIM

I think animals' features help them survive by _____.

Cite evidence from the activity.

EVIDENCE

The evidence I found in the _____
included_____.

Discuss your reasoning as a class. Tell about your discussion.

REASONING

My reasoning for my claim is _____.

You will revisit your claim to add more evidence later in this lesson.

VOCABULARY

Look for these words as you read:

adaptation

camouflage

hibernate

migrate

mimicry

Adaptations

Different beak sizes and shapes allow birds to gather food in different environments. This is an example of adaptation. An **adaptation** is a structure or behavior that helps an organism survive in its environment.

The frog uses its sticky tongue to capture an insect to eat.

Spiky edges on its leaves protect the bush from being eaten.

A polar bear has an adaptation called **camouflage**, which means it blends into its environment. Camouflage helps living things stay safe. A snake's skin pattern may match the ground it lies on, making it difficult for a predator to see the snake.

Some animals are adapted to living in cold climates. Sea lions and walruses have a layer of fat called blubber under their skin that helps them stay warm. Some animals are adapted to survive in hot temperatures. Camels have patches to protect their legs so they are not burned when the camel kneels.

1. Circle two examples in the text that show adaptations in animals.

2. Why don't all animals have the same adaptations?

_____

_____

_____

INQUIRY ACTIVITY

Hands On

Camouflaged Beans

Investigate how camouflage helps organisms hide from predators.

Make a Prediction Will black beans or black-eyed peas be harder to separate from white beans?

_____

_____

_____

Copyright © McGraw-Hill Education (2)Pauliene Wessel/123RF, (4)Ingram Publishing, (5)Ken Cavanagh/McGraw-Hill Education, (others)Jacques Cornell/McGraw-Hill Education

Materials

 3 clear plastic cups

 white beans

 black beans

 black eyed peas

 white sheet of paper

 stopwatch

Carry Out an Investigation

1. On a white sheet of paper, mix some of the white beans with some of the black beans.

2. **Record Data** Time how long it takes to separate all the black beans from the white beans.

3. On the white sheet of paper, mix some of the white beans and some of the black-eyed peas.

4. **Record Data** Time how long it takes to separate all the black-eyed peas from the white beans.

	Black Beans	Black-Eyed Peas
Time (seconds)		

Communicate Information

5. Did your results support your prediction? Explain.

_____

_____

Desert Adaptations

A desert is a very dry environment. It rarely rains in a desert. When rain does fall, it can pour down heavily. Temperatures are often very hot during the day and cold at night. Organisms that live in a desert have adaptations to help them survive in these conditions.

Water

Desert plants cannot depend on regular rainfall for their water. Instead, their roots are adapted to spread widely or grow deep to find water. A desert plant has stems adapted for storing water. Many desert animals get their water by eating plants or other animals.

Temperature

Desert animals have adaptations to keep them from being too hot during the day. Coyotes and rattlesnakes are nocturnal. This means they are active at night and sleep during the daytime. Jackrabbits stay cool by having small bodies and long ears. This helps the heat escape from their bodies. Some animals have light-colored bodies. Light colors absorb less heat.

Circle two adaptations that help desert plants live for long periods of time without water.

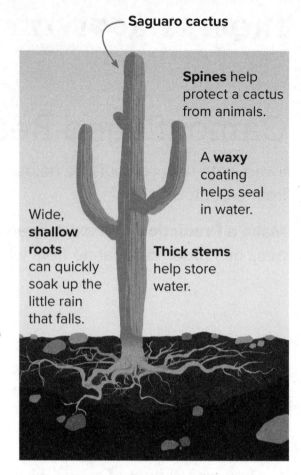

Saguaro cactus

Spines help protect a cactus from animals.

A **waxy** coating helps seal in water.

Wide, **shallow roots** can quickly soak up the little rain that falls.

Thick stems help store water.

Mesquite tree

Small leaves do not lose much water.

Thorns protect the tree from hungry and thirsty animals.

Long roots grow deep underground where they can find stored water.

Ocean

Oceans are home to millions of living things. Each living thing has adaptations that help it survive in the salty water.

⬤ **GO ONLINE** Watch the *Extreme Habitats* video to learn about extreme adaptations.

Algae

Seaweeds look like plants, but they are not. They are plant-like organisms called algae. Like plants, algae make their own food from sunlight. Most algae have structures that are like leaves. Some algae have root-like structures for attaching themselves to the ocean floor. Because they need sunlight, rooted algae can only live in shallow water. Algae that have no roots drift near the ocean's sunlit surface.

Kelp is a kind of algae. This picture shows a seaweed forest of kelp.

Ocean Animals

Whales and dolphins breathe air. They can hold their breath for a long time as they dive deep to look for food. When they need to breathe, they rise quickly to the surface. Fish, on the other hand, have gills for getting oxygen from water.

Many ocean animals have fins. Fins help them swim quickly and control their movement. Ocean animals swim for long distances when they migrate. To **migrate** means "to move from one place to another." Animals migrate to find food, to reproduce, or because the water temperature has changed.

When sperm whales migrate, they swim in groups called pods. The whales swim together for thousands of kilometers.

1. Circle two types of ocean animal adaptations.

2. Why do animals migrate?

_____

_____

REVISIT Revisit the Page Keeley Science Probe on page 23.

PAGE KEELEY
SCIENCE
PROBES

Forest

A forest is an environment in which many trees grow near one another. The tops of tall trees are in sunlight. Some organisms are adapted to life in the treetops. Others live on the ground.

GO ONLINE Explore *Animals in Their Own Environment* to see animals in different environments.

Plants

Tropical rain forests are very wet places. Too much water can harm plant leaves. Some leaves have grooves that allow water to drain off easily. Little sunlight reaches beneath the treetops. Plants living on the dim forest floor have large leaves to catch as much sunlight as possible.

A temperate forest grows where winters are cold and dry. In winter there is less sunlight for plants to make food. Many trees have adapted by shedding their leaves each autumn when the weather gets cold. Without leaves, the trees need less water.

Animals

Animals have many different adaptations for survival in forests.

Mimicry occurs when one living thing looks like another. Mimicry gives an animal a way to be hidden when not moving. It can help an organism hunt without being seen.

When a predator comes after a skunk, the skunk sprays a stinky chemical at it. A porcupine defends itself with many sharp quills.

During winter in some forests, the temperature is cold. Food is hard to find. Some animals survive by going into hibernation. **Hibernate** means to rest through winter. They use little energy, and they do not eat.

1. Why might an animal hibernate?

_____

_____

_____

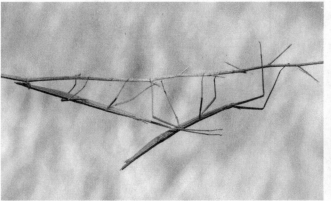

Look for the insect in this photo.

A porcupine's quills come off easily and stick into any attacking animal.

2. Look at the photos, and circle an example of mimicry.

 GO ONLINE Explore the *Rabbit Population* simulation to see how a rabbit's color affects its survival.

COLLECT EVIDENCE

Add evidence to your claim on page 29.

Cut out the Notebook Foldables tabs given to you by your teacher. Glue the anchor tabs as shown below. Use what you have learned to explain how each adaptation helps animals survive. Give an example of each adaptation.

Glue anchor tab here

Monarch   Viceroy

What Do Animal Shelter Workers Do?

Animal Shelter Workers are people who take care of animals that do not have a home. People who work in animal shelters have a lot of love for animals. They feed the animals and bathe them. They also help veterinarians treat the animals if they are sick.

Animal shelter workers also clean the animal pens and pick up homeless animals to bring them to the shelter. Animal shelter workers have an important job. They are helping animals that could not survive on their own.

It's Your Turn

Why do you think some dogs and cats are not able to survive in their environment? How might it change things if those animals moved into a shelter?

INQUIRY ACTIVITY

Hands On

Design a Bird

Design and build a model bird that is well-adapted to its environment. Use evidence to explain how the adaptations will help it survive.

Make a Prediction What adaptations does your bird need?

_____

_____

Carry Out an Investigation

1. Describe the environment where your model bird will live. Describe what the bird eats and where it finds its food.

_____

_____

_____

2. Make a sketch of your bird below. Label its adaptations and identify the materials that you will use to build it.

Copyright © McGraw-Hill Education (1 5 7)Ken Cavanagh/McGraw-Hill Education, (4)Joe Polillio/McGraw-Hill Educaiton, (6)McGraw-Hill Education, (others)Jacques Cornell/McGraw-Hill Education

Materials

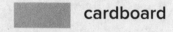 cardboard

 scissors

 masking tape

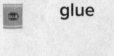 glue

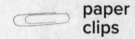 drinking straws

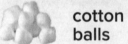 paper clips

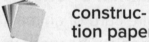 cotton balls

construction paper

3. Use classroom resources to construct your model.

Communicate Information

 What **adaptations** does your bird have that help it survive in its environment? Use **evidence** from the lesson.

_____

_____

_____

_____

_____

4. Did your results support your prediction? Explain.

_____

_____

_____

_____

_____

5. Compare your model bird with the model bird of a classmate. What are some similarities and differences between the environments and traits of both birds?

_____

_____

_____

EXPLAIN
THE PHENOMENON

How does the quail survive in its environment?

Summarize It

Explain how organisms survive in different environments.

_____

_____

_____

_____

_____

_____

REVISIT Revisit the Page Keeley Science Probe on page 23. Has your thinking changed? If so, explain how it has changed.

PAGE KEELEY
SCIENCE
PROBES

Three-Dimensional Thinking

1. Why do trees in temperate forests lose their leaves in the fall?

   A. to provide shelter for animals

   B. to conserve water

   C. to provide nutrients for their root systems

   D. to protect their roots from heavy rainfall

2. Which phrase belongs in the empty circle below?

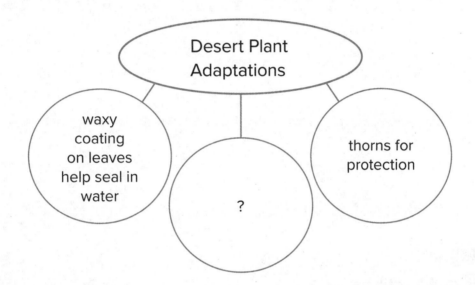

   A. few or no roots

   B. large leaves to catch sunlight

   C. grooves on leaves

   D. thick stems to store water

3. A special characteristic that helps an organism survive in its environment is a(n) _____.

   A. adaptation

   B. camouflage

   C. hibernation

   D. migration

Extend It

You are working on a school project to help increase the population of bees in your area. Research what the bees need to survive and create a plan to present your research to your classmates.

_____

_____

_____

_____

_____

_____

OPEN INQUIRY

What questions do you still have?

_____

_____

Plan and carry out an investigation to answer one of the questions.

KEEP PLANNING
STEM Module Project
Science Challenge

Now that you have learned how organisms adapt to their environment, go to your Module Project to explain how the information will affect how you design your model.

Design an Animal's Adaptations

You've been hired as an animal curator. Using what you have learned throughout this module, you will develop a design for a newly discovered animal in a specific habitat. Your goal will be to design and build a model of the animal in its habitat.

Planning after Lesson 1

Apply what you have learned about organism survival to your project planning.

What do all animals need to survive that you need to include?

Record information to help you plan your model after each lesson.

Planning after Lesson 2

Apply what you have learned about adaptations and variations to your project planning.

How does knowing about adaptations and variations affect your project planning?

Sketch Your Own Design

Draw your ideas below. Select the best one to build and test.

Design an Animal's Adaptations

Look back at the planning you did after each lesson.
Use that information to complete your final Module Project.

Build Your Model

1. Determine the materials you will need to build your model of an animal using its adaptations to survive in its habitat.

2. Use your project planning to build your model.

3. Explain how your animal's adaptations help it live in its environment.

Materials

_____

_____

_____

Communicate Your Results

You are using the Science Process!

Share the plan for your model and your results with another group. Compare the animals and their habitats. Communicate your findings below.

_____

_____

_____

_____

_____

_____

_____

_____

_____

_____

_____

_____

_____

_____

_____

MODULE WRAP-UP

REVISIT
THE PHENOMENON

Using what you learned in this module, explain how the animal survived.

> Revisit your project if you need to gather more evidence.

Have your ideas changed? Explain.

Change the Environment

How will the fire affect all the living things in the forest?

Burning Forest

⊗ GO ONLINE

Check out *Burning Forest* to see the phenomenon in action.

💬 Talk About It

Look at the photo and watch the video *Burning Forest*. What questions do you have about the phenomenon? Talk about your observations with a partner.

Did You Know?

Controlled fires can be good for a forest. They remove dead brush without killing older trees. You need permission to set a controlled fire. It takes a lot of experience to control a fire.

Past, Present, and Future

What happens to an ecosystem after a forest fire? You are being hired as a wildlife rehabilitator. At the end of the module, you will construct models of the forest at different time periods. Using what you have learned throughout this module, you will design models of the forest's past, present, and future.

Lesson 1
Fossils

Lesson 2
Changes Affect
Organisms

Wildlife rehabilitators apply their knowledge of the land and organisms to create a rehabilitation plan after a forest fire.

What do you think a wildlife rehabilitator needs to know before restoring an ecosystem?

POPPY
Park Ranger

STEM Module Project

Plan and Complete the Science Challenge You will use what you learn to make a plan to restore a forest after a fire.

Extinct Today

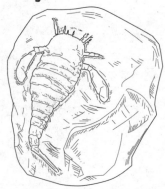

Four friends went to a science museum. They looked at the fossils of extinct plants and animals that lived millions of years ago. They had different ideas about why the plants and animals are no longer living today. This is what they said:

Norma: *I think they died out because they eventually grew too big.*

Max: *I think they just got too old and died out.*

Eshana: *I think their environment changed and they could no longer survive.*

Diego: *I think they were overhunted by humans who used them for food, clothes, and shelter.*

Which friend do you agree with the most? _____

Explain why you agree.

_____

_____

_____

You will revisit the Page Keeley Science Probe later in the lesson.

Fossils

LS4.A, LS4.C

Why don't mammoths exist anymore?

🔊 GO ONLINE

Watch *Fish Fossils* to see the phenomenon in action.

💬 Talk About It

Look at the illustration and watch the video *Fish Fossils*.
What questions and observations do you have about mammoths and the fish fossils?

_____

_____

_____

_____

_____

_____

Did You Know?

Mammoth fossils have been found from Ohio and Illinois all the way to Southern California.

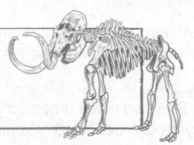

INQUIRY ACTIVITY

Hands On

Layers and Fossils, Part 1

Mammoths are no longer living, so how do we know so much about them? Earth has different layers of rock and soil. Scientists use the layers of Earth and the fossils found in them to study organisms that lived long ago.

Make a Prediction How can you tell which fossils are oldest?

_____

_____

_____

Carry Out an Investigation

BE CAREFUL Wear safety goggles to protect your eyes from the sand.

1. Use the plastic spoon to mix two tablespoons of glue with two tablespoons of water in the plastic cup.

2. Pour a thin layer of colored sand into the paper cup. The sand will represent a layer of rock.

3. Add an object to the paper cup. Cover the object with the same color of sand you used in step 2. Record the color of each layer in your model and include the "fossil" you placed there in step 6.

4. Add about 1 tablespoon of the glue and water mixture to the sand to make it stay together.

5. Repeat steps 2–4 two more times so you have three layers of "rock" and "fossils."

Materials

 safety goggles

 tablespoon

 white glue

 cup of water

 paper cup

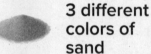

 3 different colors of sand

natural objects

Copyright © McGraw-Hill Education (1 2)Janette Beckman/McGraw-Hill Education, (3)Ken Cavanagh/McGraw-Hill Education, (4)James Turner/EyeEm/Getty Images, (5)Yanavarut Phugongchana/Shutterstock

6. Draw a diagram of what the layers should look like in the drawing box below. Label the placement of each of the fossils.

Communicate Information

7. Which "fossil" in your model was placed first?

_____

_____

8. Which fossils on Earth are the oldest? Explain.

_____

_____

9. How is it possible to find old and new fossils in the same area?

_____

_____

Talk About It

How do you think technology can help find fossils?

Fossils

A **fossil** is a trace of the remains of a living thing that died long ago. Fossils can be shells, bone, skin, leaves, or even footprints. They can form in several different ways.

Trace Fossils

Sometimes an animal leaves a footprint on the ground as it walks. The shape of the animal's footprint leaves an imprint in mud or clay. The mud hardens and changes to rock over time. These fossils are called trace fossils. They record a trace of a once-living organism.

Preserved Remains

Some fossils are the actual remains of an organism trapped in Earth's materials. Amber fossils formed when insects became trapped in tree sap, which hardened over time. Preserved remains also have been found in tar and ice.

Fossils can look exactly like the hard parts of the animals they came from.

This dinosaur footprints was left in mud. The mud hardened into rock, fossilizing the footprints.

Molds and Casts

Some fossils, such as bones and teeth, look like the actual parts of animals. When scientists find a bone of a dinosaur, they really have found rock hardened into the shape of a bone.

GO ONLINE Watch the video *Fossils* to learn more about how different fossils form.

A seashell can make an imprint, or a mark, in mud or sand. Over time, the shell is buried in more layers of sand or mud. Water and minerals seep into the ground. The shell breaks down, but the material around it hardens into rock. The shell-shaped space that is left is called a mold. The empty space can be filled with a new material to show what the original shell looked like. This kind of model is called a cast.

This insect's entire body was trapped in tree sap and is now a fossil.

1. Think back to the Inquiry Activity, *Layers and Fossils, Part 1*. What type of fossils did you create?

_____

2. How are preserved remains different from fossils in stone?

_____

_____

_____

3. What can a scientist learn from studying a mold fossil of an ancient seashell?

_____

_____

_____

_____

The fossil on the bottom is a mold. The fossil on the top is the cast.

What Fossils Tell Us

Earth is about 4.5 billion years old. People have lived on Earth only a small part of that time. Scientists learn about Earth's past by studying fossils.

We know about organisms that lived long ago because of the fossil record. Mammoths are examples of such animals. No one has ever seen a mammoth, but we know about them because people have studied their fossils.

This photo shows a fossil of a mammoth. From their fossils, scientists know they were large mammals.

Changes in Living Things

The fossil record shows that the kinds of things on Earth have changed over time. Early in Earth's history, many fish swam in the oceans. There were few animals on land. Later, many species of fish became extinct. **Extinction** is when there is no more of an organism's kind left on Earth.

Evidence of Earth's Changes

Looking at fossils also tells us about how Earth's environment has changed over time. Today, Antarctica is a cold place. It is covered in snow and ice. Scientists have found fossils of leaves and wood there. These fossils tell scientists that Antarctica was once a warm, wet area.

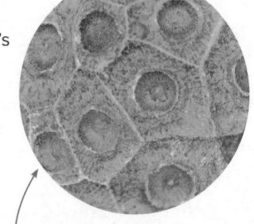

Scientists use details they find from fossils to piece together the story of Earth's past. From each new fossil found in Earth's surface, we learn a little more about our planet's history.

This coral once lived in a warm sea. Over millions of years, the area has changed to dry land in the center of the continent.

REVISIT Revisit the Page Keeley Science Probe on page 51.

PAGE KEELEY
SCIENCE PROBES

INQUIRY ACTIVITY

Hands On

Layers and Fossils, Part 2

Uncover the "fossils" in the layers of rock that were made earlier in this lesson.

State the Claim How are your "fossil" cups like real fossils that scientists discover?

Materials

safety goggle

"fossil" cups

paper plate

paintbrush

_____

_____

Make a Model

1. Trade cups with another group.

2. Place the cup on the plate and carefully remove the rock formation from the cup.

3. Starting at the top, brush away the sand or use the end of a brush to uncover the fossils.

4. **Record Data** On a separate sheet of paper record the objects found and the order.

Communicate Information

5. Did your findings support your claim? Explain.

_____

_____

 You made a **model** of part of Earth's system. How is your model like Earth? How is it different?

_____

_____

_____

Learning from Fossils

Some fossils give clues about a living thing's size, shape, and environment. The woolly mammoth became extinct thousands of years ago. Fossils tell us that it had a large trunk and tusks. Yet, fossils cannot tell us how this animal used its body parts. Instead, scientists learn how animals that lived long ago used their bodyparts by studying similar animals living today.

What do elephants and wooly mammoths have in common? What is different about them?

How deep a fossil is buried gives clues about when an organism lived. Fossils found closest to the surface are usually the youngest. Fossils found in deeper layers are usually older. This helps scientists put together a picture of what organisms lived at the same time.

Clues in rocks can also give hints about what Earth was like at different points in the past. Many fish fossils are found on land. This means that millions of years ago, that land was covered with water. Over time, the land rose above the water. Fossils remained in the rock and soil that had been underwater.

Scientists may find many fossils buried at the same depth. What do they know about these fossils?

Simulation

Fossil Dig

⊙ GO ONLINE

Use *Fossil Dig* simulation to investigate the layers of a fossil dig.

Make a Prediction What information about an organism can you get from a fossil dig?

_____

_____

Choose one area of the simulation and record data below.

Layer	Did it live on water or land?
A	
B	
C	
D	

Communicate Information

1. How did you know whether an organism lived on land or in water?

_____

_____

2. How did you know which fossils were older?

_____

3. Select one fossil and describe the environment in which it lived.

_____

FOLDABLES®

Cut out the Notebook Foldables tabs given to you by your teacher.
Glue the anchor tabs as shown below. Describe the fossils and
animals shown in the pictures and explain what they tell us about
prehistoric life.

Glue anchor tab here

Glue anchor tab here

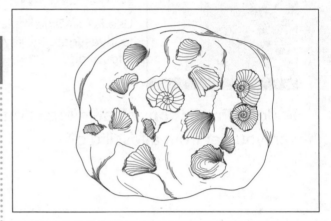

What Does an Archaeologist Do?

Archaeologists study human history. They conduct digs to uncover fossils and the tools humans used in the past. They learn about past human societies and how humans have developed and changed over time.

Francis Turville-Petre was an archaeologist. He worked in different parts of the world conducting digs. He is famous for discovering a new fossil never seen before.

It's Your Turn

Think like an archaeologist. Complete the activity on the next page to uncover the answers that fossils hold.

💬 Talk About It

What would you be most interested in researching about human life in the past?

INQUIRY ACTIVITY

Hands On

Fossil Mystery

Make a model of a fossil, and see whether your classmate can identify your fossil. Use the fossil to determine the environment the animal lived in.

Make a Prediction What can you learn about a fossil from its characteristics?

_____

Carry Out an Investigation

Materials

 colored pencils

paper

1. Choose a favorite animal. Then use a pencil and the key to draw your fossil on a separate sheet of paper.

If your animal is a...	then draw a...
mammal	circle
bird	square
amphibian	rectangle
reptile	triangle
fish	star

2. Use the key below to draw marks on your fossil.

If your animal...	then mark your drawing with...
only lives in water	red triangle
only lives on land	blue circle
lives in water and on land	purple star
is a carnivore (eats meat)	black square
is an herbivore (eats plants)	yellow square
is an omnivore (eats meat and plants)	green square

3. Trade your model fossil with a person sitting next to you. Use the key to find out about the animal your classmate chose.

4. Using what you have learned about the mystery fossil, draw a picture of the animal the mystery fossil represents. In your drawing show the enviornment the animal lived in and what it ate for food.

Communicate Information

5. Share your drawing with a partner. Did your drawing represent the fossilized animal they chose? Explain.

_____

_____

_____

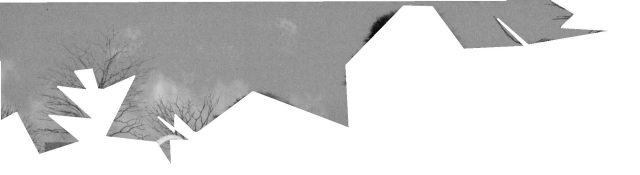

EXPLAIN
THE PHENOMENON

Why don't mammoths exist anymore?

Summarize It

Explain what fossils tell us about the environment.

_____

_____

_____

_____

_____

_____

REVISIT **PAGE KEELEY SCIENCE PROBES** Revisit the Page Keeley Science Probe on page 51. Has your thinking changed? If so, explain how it has changed.

Three-Dimensional Thinking

1. A type of organism that has no living population is said to be _____ .

   A. fossil

   B. remains

   C. an organism

   D. extinct

Use the table below to answer question 2.

Animal	Feature
saber-toothed cat	long, sharp teeth
woolly mammoth	flat teeth
pterodactyl	long wingspan
triceratops	leathery skin

2. Which animal was most likely a meat-eater?

   A. saber-toothed cat

   B. woolly mammoth

   C. pterodactyl

   D. triceratops

3. The fossil of a fish was found at the top of a mountain. Which statement is MOST likely true?

   A. The mountain was once a hill.

   B. Fish used to live on the mountain.

   C. The mountain was once under water.

   D. Someone moved the fossil to the mountain.

Extend It

You have been asked to be a guest teacher in a first grade classroom. You will need to create an activity to do with your students about extinct animals. Be creative when planning your activity. Design a game, write a skit to perform, or create a craft project to help teach your class.

_____

_____

_____

_____

_____

_____

OPEN INQUIRY

What questions do you still have?

_____

_____

Plan and carry out an investigation to answer one of the questions.

KEEP PLANNING
STEM Module Project
Science Challenge

Now that you have learned about fossils, go to your Module Project and explain how the information will affect your plan for your model.

Changes in Ecosystems

Changes in ecosystems affect organisms. Some of these changes include the temperature, food and water supply, and shelter an organism needs to survive. Circle any of the boxes that best describe what can happen to a group of organisms when there is a forest fire.

None of the organisms will survive and reproduce.	Some organisms will survive and reproduce.	All the organisms will survive and reproduce.
None of the organisms will move to new locations.	Some of the organisms will move to new locations.	All of the organisms will move to new locations.
None of the organisms will move into the changed environment.	Some of the organisms will move into the changed environment.	All of the organisms will move into the changed environment.
None of the organisms will die.	Some of the organisms will die.	All of the organisms will die.

Explain your thinking. How did you decide what happens to organisms when there is a forest fire?

_____

_____

_____

You will revisit the Page Keeley Science Probe later in the lesson.

Changes Affect Organisms

ENCOUNTER
THE PHENOMENON

How will the broken dam affect the area around it?

Oroville Dam Breach

▶ GO ONLINE

Check out *Oroville Dam Breach* to see the phenomenon in action.

💬 Talk About It

Look at the photo and watch the *Oroville Dam Breach* video. What observations and questions do you have about the phenomenon?

Did You Know?

In February 2017, the Oroville Dam's main spillway was damaged. Water gushed over the emergency spillway. Experts feared the dam would collapse. Almost 200,000 people were evacuated.

Copyright © McGraw-Hill Education

INQUIRY ACTIVITY

Changes by Humans

You used fossils to investigate how Earth has changed over millions of years. Many things change suddenly, like when a dam breaks. Investigate how humans have changed the environment.

Make a Prediction How do you think humans have changed the environment around your school?

_____

_____

Carry Out an Investigation

1. Go outside and record your observations about the environment around your school.

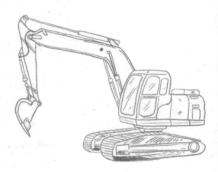

2. On a separate piece of paper, record the living things, nonliving things, and the things made by humans that you see.

Communicate Information

3. Look at the things that you listed as living, nonliving, and human-made. Describe how humans have changed your environment.

_____

_____

_____

 **Talk About It**

Did your findings support your prediction? Explain.

MAKE YOUR CLAIM

How have humans changed their environment?

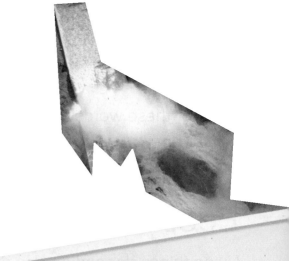

Make your claim. Use your investigation.

CLAIM

Humans change their environment by _____.

Cite evidence from the activity.

EVIDENCE

The investigation showed that _____.

Discuss your reasoning as a class. Tell about your discussion.

REASONING

The evidence that supports the claim is _____.

You will revisit your claim to add more evidence later in this lesson.

Ecosystems

Living things depend on one another. They also depend on nonliving things, such as sunlight. Earth has many different kinds of ecosystems. Different organisms live in different parts of an ecosystem. Living things get food, water, and shelter from their habitats. Many different habitats make up an ecosystem.

Interactions in Ecosystems

The different parts of an ecosystem interact. All of the parts depend on and affect one another. For example, animals depend on plants for food and shelter. Squirrels get acorns from trees to eat and gather branches to make nests in trees. Squirrels can change and affect how trees grow. Trees and other plants give off oxygen that animals need to survive.

Living things also depend on and affect nonliving things in an ecosystem. The grass on a meadow could not grow without air, water, sunlight, nutrients, and soil. When chipmunks and rabbits dig in the ground, they break up rocks, which helps form the soil.

Seeds blow onto bare ground. The environment changes as plants take in water and nutrients.

As more plants grow, animals move in to the environment. They use the plants for food and shelter.

NPS Photo/Cookie Ballou

Copyright © McGraw-Hill Education

Bison and other animals in this meadow depend on grasses for food. The grasses need air, water, sunlight, and soil to grow.

GO ONLINE Watch the video *How Environments Change* to learn more about different types of changes.

Changes in Ecosystems

Every living thing changes its ecosystem as it meets its needs. A spider spins a web to catch insects for food. A cotton plant takes water from the soil. Organisms reproduce and grow in number. These events change an ecosystem in small ways.

Other living things make bigger changes to their ecosystems. For example, bacteria, worms, and mushrooms break down leaves and dead animals. This returns valuable nutrients to the soil. Later, plants can use those nutrients to grow.

In time the plants grow larger. They compete for water, space, and sunlight. Animals compete for food and water.

Trees block sunlight from reaching smaller plants. These plants may die as trees grow larger.

Copyright © McGraw-Hill Education   U.S. Fish & Wildlife Service

People change ecosystems more than other organisms do. Some changes, such as planting trees, are helpful. Others, such as draining wetlands to build over them, hurt ecosystems.

The arrival of a new type of organism can cause either big or small changes to an ecosystem. An **invasive species** is an organism that is introduced into a new ecosystem either by accident or on purpose by humans. Without predators, the population of the invasive species increases rapidly. Invasive species can be plants or animals.

Natural events can cause dramatic changes in ecosystems. Droughts may cause an entire population of animals to leave an area in search of food sources. Floods and fires can sweep away plants and animals. It can take years for an ecosystem to recover from these changes.

COLLECT EVIDENCE

Add evidence to your claim on page 73 about how humans change their environment.

A North American beaver works on its dam in Grand Teton National Park, Wyoming.

Invasive wild pigs are digging for food near Cape Canaveral, Florida.

How might a beaver dam change the ecosystem?

🌀 **GO ONLINE** Explore *Environments: Before and After* to see the effects of environmental changes.

Historical Dam

WRITING ▸ **Connection** Look at the two photos below. On a separate sheet of paper, write a paragraph describing three ways the ecosystem changed when the dam gates opened. Also, describe a possible problem that could have occurred if the dam was not built.

PRIMARY SOURCE

Dam 96 can be found at the J. Clark Salyer Wild Refuge.

After Dam 96 at the Souris River Basin was opened.

Inspect

Read the passage *Southwest Forest Fires*. Underline text evidence about how a park ranger helps with forest fires.

Find Evidence

Reread What happens to animals after a fire? Highlight evidence that supports your answer.

Notes

_____

_____

_____

_____

_____

_____

_____

_____

It takes a lot of firefighters working together to put out wildfires.

Southwest Forest Fires

The southwest region of the United States can experience wildfires when conditions are dry due to lack of rain. This area of the country often experiences periods of drought. Once a fire starts, it is hard to stop it from spreading. This is especially true when a forest is dry.

Park rangers have an important job when there is a wildfire. First, they let people know to stay away. They do this by talking to news reporters and by using social media. After the fire is out, they take care of the animals that survived the fire. Some small animals hide underground or in logs. Park rangers help to rehabilitate the animals.

Make Connections

💬 Talk About It

Why might it be difficult for humans and wild animals to live in the same habitat?

Notes

_____

_____

_____

_____

_____

_____

_____

_____

_____

_____

Experts say birds and large animals usually survive wildfires. Flying above the smoke and flames or large strides make them able to flee quickly when a fire first begins.

Where do you think wild animals go after their homes are destroyed? Most animals find another forest to call home. It can take time to find a new forest. This is why some people who live in areas that experience forest fires see wild animals like coyotes walking along the highway. It is not safe for wild animals to explore outside of their natural habitat, though. They are not used to living among humans. As animals seek refuge, they sometimes wind up in neighborhoods. In these cases, park rangers and other wildlife experts are called in to help with their relocation.

REVISIT Revisit the Page Keeley Science Probe on page 69.

What Does a Park Ranger Do?

Park Rangers work in state or national parks. They lead tours and give visitors information about the parks. They spend a lot of time outside, and they sometimes work inside a visitor center.

That sounds like fun, but park rangers also have to do a lot of hard work. Sometimes they need to find lost hikers. In some parts of the country, park rangers need to protect people and animals if there is a forest fire.

It's Your Turn

Think like a park ranger. Complete the activity on the next page to explore how a park ranger might work to solve problems that arise.

💬 Talk About It

As a park ranger, what could you do to protect animals and park visitors if there were a forest fire? How might a forest fire change the environment in a park?

Research

Solve for an Invasive Species

A park ranger has noticed that an invasive species has taken over. Research an invasive species, define a problem it causes, and design a possible solution to help the park ranger.

Research Find information about an invasive species that interests you. Record your notes on a separate piece of paper.

1. What is the name and location of the invasive species?

2. How did this species affect the ecosystem it is in now?

State a Claim Identify a problem caused by the invasive species.

_____

_____

Design a Solution

3. Use what you have learned about ecosystems to design a model of a solution to the problem caused by the invasive species that you researched.

Communicate Information

 Talk About It

4. Share your solution with a partner. Ask your partner how they would build on or improve your idea. Write the ideas above in your solution.

EXPLAIN
THE PHENOMENON

How will the broken dam affect the area around it?

Summarize It

Explain how humans affect their environment.

_____

_____

_____

_____

_____

_____

REVISIT Revisit the Page Keeley Science Probe on page 69. Has your thinking changed? If so, explain how it has changed.

Three-Dimensional Thinking

1. What might happen if you introduce new plants or animals to an ecosystem?

_____

_____

_____

_____

_____

_____

_____

2. Explain how an invasive species and humans cutting down trees in the rainforest are similar.

_____

_____

_____

_____

_____

Extend It

Create a skit. The skit should be about someone that wants to plant a fern that is known to be invasive in a community garden. Inform them of the risks and come up with a solution.

_____

_____

_____

_____

_____

_____

_____

_____

_____

_____

KEEP PLANNING

STEM Module Project
Science Challenge

Now that you have learned about how changes affect organisms, go to your Module Project to explain how the information will affect your plan for your forest.

Past, Present, and Future

You have been hired as a wildlife rehabilitator. You will construct models of the forest at different time periods. Using what you have learned throughout this module, you will design models of the forest's past, present, and future.

Planning after Lesson 1

Apply what you have learned about fossils that will help in your project planning.

How can you use fossils to create a model of the forest?

Record information to help you plan your model after each lesson.

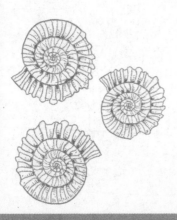

Poppy
Park Ranger

Planning after Lesson 2

Apply what you have learned about changes in the environment to your project planning.

How does a change in the environment cause problems for plants and animals? Give an example.

Sketch Your Model

Draw your ideas of what your forest looked like before a forest fire. In your sketch, include living and nonliving things.

Research

🔍 Read the Investigator article *Wildfires Renew Forests* to learn more about how forest fires affect the environment. Using the resources provided by your teacher, research how a forest is restored after a fire. List living and nonliving things that will be in your plot of land in the past column. You will need to list how the living and nonliving things will be restored.

Past	Future

Past, Present, and Future

Look back at the planning you did after each lesson. Use that information to complete your final module project.

Build Your Model

Materials

1. Make a plan for how you will restore the forest.

2. Make a list of living and nonliving things, including fossils, found in the forest.

3. Design a model of the forest before the fire, immediately after the fire, and four years after the fire.

4. List the materials you will need to build your model.

5. Share your plan and designs with a classmate and discuss if your plan will work.

6. Record the improvements made to the design after meeting with your classmate.

Procedure:

_____

_____

_____

Design Your Solution

Use the space below to draw your final
models. Include labels.

You are using
the Scientific Process!

Communicate Your Results

Share the plan to restore your ecosystem with another group. Compare and discuss changes that could be made to improve on your restoration.

Use what you have learned in this module and through your research. Communicate your findings below, and explain how you would improve your original plan.

MODULE WRAP-UP

REVISIT
THE PHENOMENON

Using what you learned in this module, explain how the fire affects all living things in a forest.

Revisit your project if you need to gather more evidence.

Have your ideas changed? Explain.

_____

_____

A

adaptation a structure or behavior that helps an organism survive in its environment

atmosphere a blanket of gases and tiny bits of dust that surround Earth

attract to pull toward

axis an imaginary line through Earth from the North Pole to the South Pole

B

balanced forces forces that cancel each other out when acting together on an object

birth the beginning or origin of a plant or animal

C

camouflage an adaptation that allows an organism to blend into its environment

climate the pattern of weather at a certain place over a long period of time

competition the struggle among organisms for water, food, or other resources

D

direction the path on which something is moving

distance how far one object or place is from another

E

ecosystem the living and nonliving things that interact in an environment

electrical charge the property of matter that causes electricity

environmental trait a trait that is affected by the environment

extinction the death of all of one type of living thing

F

floodwall a wall built to reduce or prevent flooding in an area

force a push or pull

fossil the trace of remains of living thing that died long ago

friction a force between two moving objects that slows them down

G

germinate to begin to grow from a seed to a young plant

group a number of living things having some natural relationship

H

hibernation to rest or go into a deep sleep through the cold winter

I

inherited trait a trait that can be passed from parents to offspring

instinct a way of acting that an animal does not have to learn

invasive species an organism that is introduced into a new ecosystem

L

learned trait a new skill gained over time

levee a wall built along the sides of rivers and other bodies of water to prevent them from overflowing

life cycle how a certain kind of organism grows and reproduces

lightning rod a metal bar that safely directs lightning into the ground

M

magnet an object that can attract objects made of iron, cobalt, steel, and nickel

magnetic field the area around a magnet where its force can attract or repel

magnetism the ability of an object to push or pull on another object that has the magnetic property

metamorphosis the process in which an animal changes shape

migrate to move from one place to another

mimicry an adaptation in which one kind of organism looks like another kind in color and shape

motion a change in an object's position

N

natural hazard a natural event such as a flood, earthquake, or hurricane that causes great damage

P

pole one of two ends of a magnet where the magnetic force is strongest

pollination the transfer of pollen from the male parts of one flower to the female parts of another flower

population all the members of a group of one type of organism in the same place

position the location of an object

precipitation water that falls to the ground from the atmosphere

R

repel to push away

reproduce to make more of their own kind

resource a material or object that a living thing uses to survive

S

season one of the four parts of the year with different weather patterns

static electricity the build up of an electrical charge on a material

survive to stay alive

speed a measure of how fast or slow an object moves

T

temperature a measure of how hot or cold something is

trait a feature of a living thing

U

unbalanced forces forces that do not cancel each other out and that cause an object to change its motion

V

variation an inherited trait that makes an individual different from other members of the same family

W

weather what the air is like at a certain time and place

✂ - - - cut on all dashed lines - - -   ▭ fold on all solid lines

To **mimic** is to look or act like someone or something else.

_____ is an adaptation in which one kind of organism looks like another kind in color and shape.

_____ means "to move from one place to another." Therefore,

_____ means "moving from one place to another."

_____ means "to rest or go into a deep sleep through the cold winter." So,

_____ means "resting or going into a deep sleep through the cold winter."

mimic

migration

hibernation

Dinah Zike's
**Visual
Kinesthetic
Vocabulary** ®

✂ cut on all dashed lines

fold on all solid lines

e

e

ry

Memory Maker: A chameleon is an animal that is known for its **mimicry**. How does a chameleon mimic other organisms?

_____

_____

_____

_____

Memory Maker: Below is a list of animals. Circle the names of animals that practice **migration**.

goose   **cat**   **salmon**   **butterfly**   **earthworm**   **sheep**

Memory Maker: Below is a list of animals. Circle the names of animals that practice **hibernation**.

tiger   **squirrel**   **bear**   **lion**   **chipmunk**   **monkey**   **bat**

Dinah Zike's
V K V
Visual
Kinesthetic
Vocabulary®

✂ cut on all dashed lines   ▭ fold on all solid lines

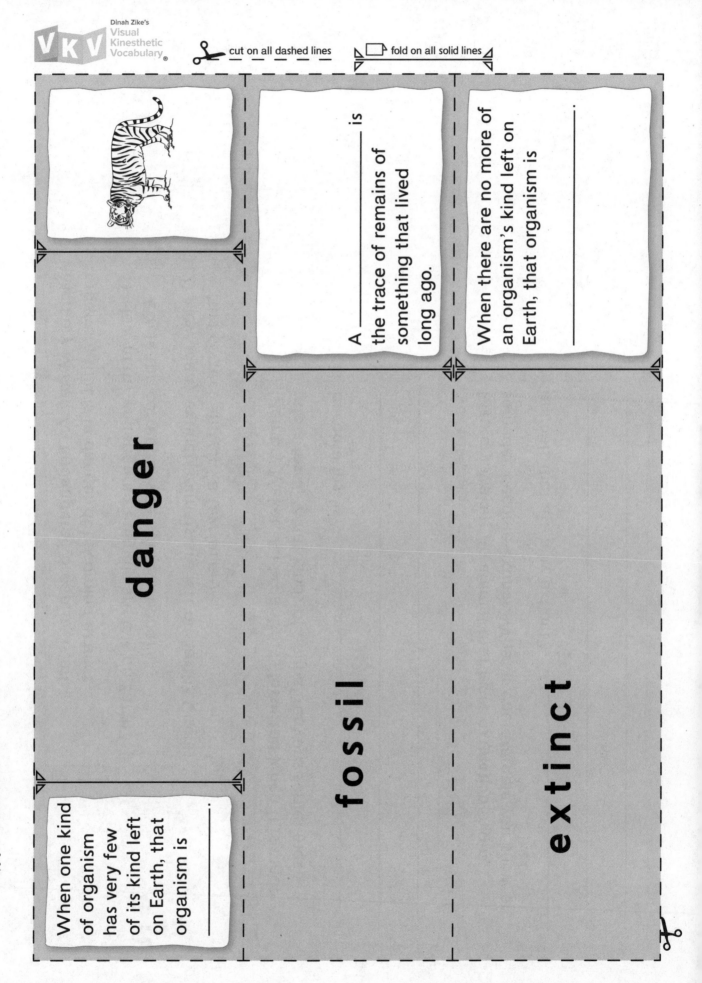

danger

fossil

extinct

A _____ is the trace of remains of something that lived long ago.

When there are no more of an organism's kind left on Earth, that organism is _____.

When one kind of organism has very few of its kind left on Earth, that organism is _____.

ion

ized

ed

Memory Maker: An **endangered** organism is an organism that is in **danger** of becoming extinct.

1) How many **endangered** animals are left on Earth? Circle the correct answer. (a few/none)

2) How many extinct animals are left on Earth? Circle the correct answer. (a few/none)

Memory Maker: An organism is **fossilized** when it is made into a **fossil**. What kinds of animal remains sometimes become fossils? _____

Memory Maker: An animal that faces **extinction** might someday become **extinct**. What is one example you know of animals that are now extinct? _____

en

VKV4 Module: Change the Environment